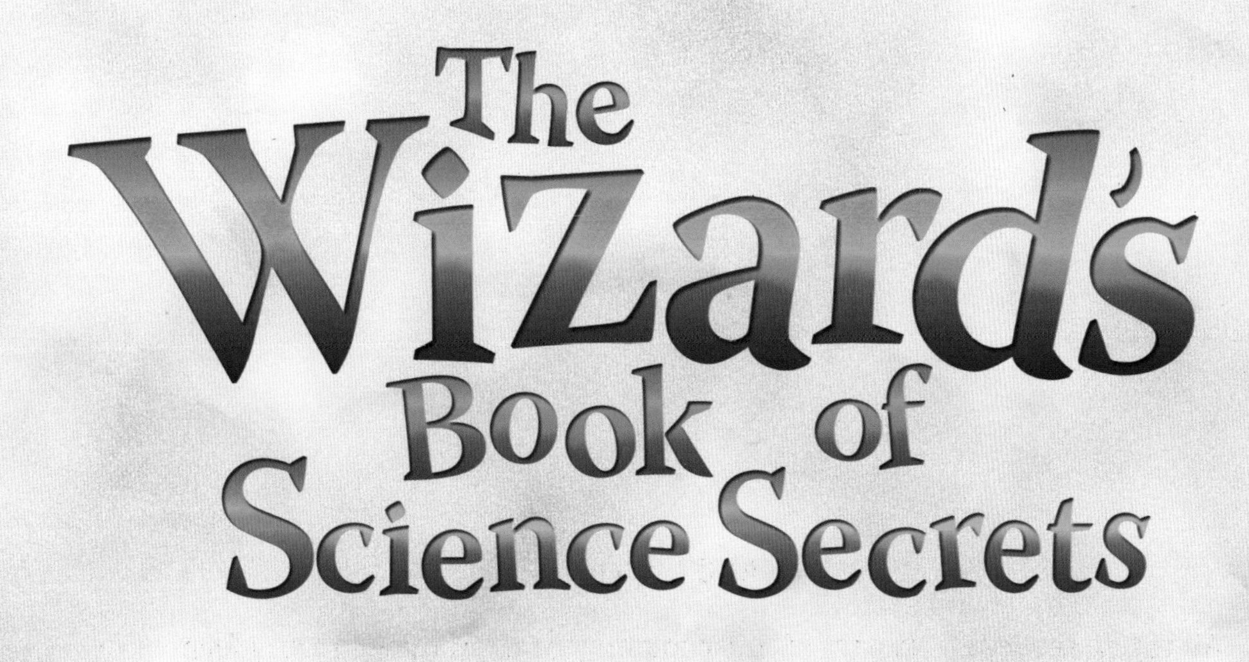

The Wizard's Book of Science Secrets

ISBN 978-0-9836066-1-1

Design and production
Lawrence Art & Design Studios
Lead design: Donavan Lawrence

Illustrations
Donavan Lawrence
iStock.com

HiZinc, LLC
333 Kilauea Avenue, Suite 214
Hilo, Hawaii 96720

Printed in Korea

Welcome back!

This is the second book in a seven-book series named after the well known, imaginary character in physics-learning lore, Roy G. Biv. His name is a learning tool – an acronym – for remembering the seven basic colors of the visible light spectrum: red, orange, yellow, green, blue, indigo and violet.

In their work, scientists often use acronyms on a daily basis. They also use a set of skills often referred to as science process skills. It does not matter in what field a scientist works.... geology, chemistry, physics, biology, etc.; all scientists master a set of basic skills.

The activities within this book are fun. But each one also incorporates at least one science process skill.... measuring, predicting, observing, analyzing, etc. A young scientist will sharpen his or her scientific thinking, while using hardware store materials to craft all sorts of activities and experiments.

In the early 19th century, a young scientist named Michael Faraday used simple items to create the world's first electromagnet. His work with it led him to discoveries that changed the world. You can find some of the same items he used in, of all places, a hardware store!

Safety

Follow these safety rules when performing activities in this book:

1. Wear eye protection when doing a science activity.

2. Learn how to safely dispose of all materials used in the activity.

3. Keep a fire extinguisher and first-aid kit handy, and know how to use them.

4. Wear appropriate clothing when performing any activity. Avoid wearing loose or bulky clothing.

5. Understand all elements of the activity before performing it. Ask questions if you do not understand.

6. Use only the size of equipment and quantities of materials suggested in the directions.

7. Make certain all people and property in the area of your science activity are well protected and informed.

Delve once again into the world's largest collection of science activities. For over 200 years, a lineage of wizards has perfected the art of engaging young minds in the skills of scientific thinking. Using common, everyday items, you can make scientific discoveries while using skills employed by scientists around the world. This book contains 25 activities requiring materials found in the local hardware store, a place all young scientists love to visit.

Contents

1. Eerie Lights 9
2. Resonance Pipes 11
3. Bobby-Pin Metallurgy 13
4. Metal-Washer Gyroscope 15
5. Inertia Salad 17
6. Piezo Poppers – Hand-Held Cannons 19
7. Internal Reflection 21
8. Gravity Strip 23
9. Hot Hands 25
10. Tasty Candles 27
11. Two-Temperature Water 29
12. Vacuum-Packed Friends 31
13. Floating Pennies 33
14. Microwave Light Bulbs 35
15. Puffed-Rice Fleas 37
16. Gremlins in a Can 39
17. Fire Tornado 41
18. Salt Engineering 43
19. Bubble Belly Dancers 45
20. A Haunted Bathroom 47
21. Gelatin-Optic Fibers 49
22. Sandwich-Bag Dartboard 51
23. Tree Drinking 53
24. Chain-Loop Ecology Hunt 55
25. Impossible Rope Trick 57

All of these activities came from the collections of four Wizards. Some were presented over 200 years ago. Over the centuries they have inspired millions of young wizards and scientists to enjoy the process of scientific thought while learning and having fun.

Eerie Lights

What is it? Without using the normal source of electricity, you can make a light bulb glow, making eerie pulses of light!

What you'll need:

Checklist:

- ❑ A fluorescent light bulb; any size, any shape
- ❑ Sheet of cellophane or plastic sandwich wrap
- ❑ A very dark room

Important Notes:
- Wear safety glasses.
- Let your eyes adjust to the darkness before starting the activity.
- This will not work very well if the air is humid. Try it on a dry day.
- Give yourself elbow room

Here's How:

1. Firmly grip a fluorescent light bulb with one hand. Use the other hand to stroke lengthwise, in one direction against the body of the bulb, a piece of clear plastic wrap. Long, slightly firm strokes work best. The phosphorescent coating on the inner surface of the bulb will emit light near the area of contact with the plastic wrap.
2. Try different stroking materials to see which works best.

The Science:

The interior surface of the glass tube is coated with a phosphorescent material, usually in the form of a white powder. When you stroke the bulb, electrical charges are generated. A magnetic field is created around the area of electrical charge, which induces the phosphors to emit light. Note that the light is emitted only when the charging cloth is in motion. A magnetic field is produced when electrons are in motion.

Sketches & Observations:

Notes:

Resonance Pipes

What is it? Make an ordinary pipe sing...
Or even roar!

What you'll need:

Checklist:

- ❑ Metal pipe, two inches or greater in diameter
- ❑ Heavy wire screen
- ❑ Propane torch or Bunsen burner
- ❑ Oven mitts

Important Notes:
-Prepare a safe surface to rest the hot pipes.
-Keep flammable materials away from the pipe and torch including you!
-Wear oven mitts.

Something to ponder:
Why is the sound created only when the pipe is held upright?

Here's How:

1. Obtain a two- or three-foot length of pipe. The length does not need to be exact. You may want to obtain pipes of assorted lengths to create a variety of tones. Wad up a piece of heavy wire mesh and stuff it into one end of the pipe. The mesh wad should be three or four layers thick.
2. Holding the tube vertically with the wire mesh at the lower end, use a torch to heat the mesh red hot. When the tube is removed from the heat source, a resonating tone should be emitted from the tube.
3. Tilting the tube to a horizontal position will cause the tone to cease. Returning it to a vertical position will cause the tone to return.

The Science:

Heat from the screen mesh warms surrounding air. An updraft is created as warm air rises in the tube. Turbulence in the rising air flow generates a tone with a wave length corresponding to the length of the tube. As the air flow continues, additional waves of the same frequency are generated, resulting in the resonating tone emitted from the tube.

Longer tubes generate waves of correspondingly longer wavelengths, thus emitting tones with lower pitches.

Sketches & Observations:

Notes:

Bobby-Pin Metallurgy

What is it? Use hairpins to create a miniature blacksmith shop. Learn the science secrets of sword making.

What you'll need:

Checklist:

- ❑ Several metal bobby pins
- ❑ Tongs
- ❑ Metal or heat-proof pad
- ❑ Safety gloves and goggles
- ❑ Propane torch or Bunsen burner
- ❑ Shallow bowl of cool water

Note:
- Hardened pins are not flexible and will snap when students attempt to bend them.
- Tempered pins show characteristics of both annealed and hardened pins.

Note:
- Annealed pins bend much easier.

Here's How:

1. Use tongs that will withstand high temperatures. Wear eye protection.
2. The high heat of the burner flame will be used to demonstrate three heat-treating procedures: annealing, hardening and tempering.
3. Annealing: Rapidly heat a bobby pin until it is red hot. Remove it from the flame and allow it to cool slowly in air.
4. Hardening: Rapidly heat a second bobby pin until it is red hot. Remove it from the flame and cool it rapidly by dropping it in a bowl of cold water.
5. Tempering: Harden a third bobby pin as described above. Return the hardened pin to the burner flame, but this time, heat it slowly. Do not heat it until it is red hot. Heat it in a less-hot region of the flame. Allow it to cool slowly in air.

The Science:

Applying heat causes metal atoms to move about faster. The atoms become more disorganized as more heat is applied. During the slow-cooling portion of the annealing process, fast-moving atoms slow down at a rate that allows them to nest together in a more organized fashion, creating a more ordered structure. The orderly crystalline structure allows metal atoms to more easily slide over each other, making the metal more flexible. Annealing is described as making metal soft. Hardened metal is created when the hot metal is quickly cooled and its disorganized atoms are locked in place. Tempered metal has a bit of both characteristics.

Something to ponder:

How have the color and flexibility of the three pins changed? For what could each type of heat-treated metal be used?

Sketches & Observations:

Notes:

METAL - WASHER GYROSCOPE

What is it? Build the world's least expensive operating gyroscope.

What you'll need:

Checklist:

- ❑ Rubber balloons
- ❑ Metal hardware, washers of different sizes and/or coins
- ❑ Safety glasses

- Most airplanes have a gyroscope that works on the same principle as the balloon. It indicates how level the plane is flying.

Research: Isaac Newton, inertia

Hints & notes:
- Transparent or light-colored balloons work best for this activity.
- Inflate the balloon until it just fits into your cupped palm.
- Don't let the balloon flop about, grasp it with your palm.

Here's How:

1. Although this activity is relatively harmless, there might be an occasion for washers or pieces of rubber to fly uncontrolled across the room. Eye protection should be employed.
2. Insert a penny into a deflated balloon. Inflate the balloon and tie its neck shut. Inflate the balloon until it comfortably fits in the palm of your own hand. Bigger is not better. Over-inflation will cause you to make deep finger indentations as you attempt to grip the balloon.
3. Gently hold the balloon in one hand. With a swirling action, achieve what might appear to be a remarkable feat: Make the penny roll on its edge around the inner surface of the balloon.
4. With practice, you'll soon be able to get the penny rolling on its edge with a simple flick of your wrist. Once you are a master at balloon penny rolling, try to get the penny rolling as fast as you can. Try to keep the penny traveling a distinct planar path.

The Science:

One of the basic laws of physics as described by Newton suggests that an object in straight-line motion tends to stay in motion until acted upon by some outside force. You might comment that the coin is not traveling in a straight line, but rather a circle.

At any given moment the coin is attempting to travel in a straight line. Popping the balloon with a pin at any time the coin is revolving will prove that point. The cause of the tendency of a rotating body to travel in a straight line is often referred to as centrifugal force. When the coin is in circular motion, you should be able to feel the effect of centrifugal force as you try to slowly rotate the balloon while the penny is in motion.

Once you have the penny traveling in a flat rotational plane inside the balloon, slowly rotate the balloon and make observations of the path of the penny. The penny should remain in its initial plane of rotation, even as the balloon rotates out of position.

Track the motion of the penny as it eventually spirals to the point it drops over. Why does the penny eventually slow down?

Sketches & Observations:

Notes:

Inertia Salad

What is it? It's a mess! But it's the slickest way to explosively slice fruits and vegetables.... using science!

What you'll need:

Checklist:

- ❑ A wood frame; approx. two feet square
- ❑ Approx. 75 feet of strong, thin wire similar to single-strand piano wire
- ❑ Plastic paint tarp
- ❑ Nails or screws
- ❑ An assortment of fruits and/or vegetables

Here's How:

1. Fabricate a wood frame from 2x4 lumber. It must be well built to withstand the tension of attached wires. Use plenty of nails or screws. If it is not appropriate to use C-clamps to attach the frame to a table or bench, add support legs so the frame can be placed upright and stable.
2. Use nails or screws to create a one-inch-square wire grid. For strength, weave the wires in an alternating, over-under pattern. Make certain the wires are taut and well attached to the frame.
3. To aid in cleaning up, you may want to spread a plastic paint tarp on the floor before starting this activity. With the screen frame sitting upright and securely placed, throw an apple or other vegetable directly at the wire grid. The apple should pass through the screen. In the process, the apple will be cut into squares.
4. All types of vegetables can be diced in this fashion. An entire head of lettuce, cantaloupe, squash, potatoes, celery; all succumb to the dicing action of the wire grid.

The Science:

If a you attempt to push your knuckle into a potato, you more than likely will not be able to break through the potato skin. However, you can easily puncture the potato skin with a fingernail. The same amount of force applied across a smaller surface area creates a greater ratio of force to area.

Even vegetables must obey the laws of motion. A tomato in motion will tend to stay in motion until acted upon by some outside force. The portion of the tomato that contacts the wire in the grid tends to stop. The remaining portion of the tomato tends to keep on moving.

The fast-moving vegetable coming in contact with the taut wire is easily cut because the pressure at the point of contact with the wire can exceed thousands of pounds per square inch.

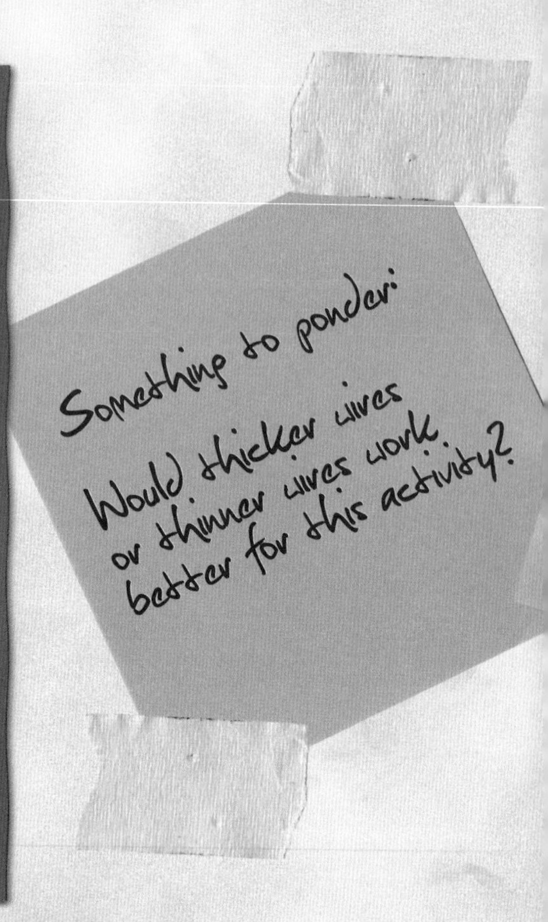

Sketches & Observations:

Notes:

PIEZO POPPERS - HAND-HELD CANNONS

What is it? Tiny explosions are safely created in plastic film canisters.

What you'll need:

Checklist:

- ❑ Common sense
- ❑ Plastic film canister with lid
- ❑ Electric lamp cord
- ❑ Methyl alcohol
- ❑ Dropper pipette
- ❑ Electrical tape
- ❑ Nail or sharp awl
- ❑ Push-button type gas grill igniter

Caution:
- Use only alcohol!
- And only the suggested amount!!

Most hardware stores sell replacement igniters for natural gas or propane grills. Usually priced at $2 to $3; any brand will suffice for this activity.

Here's How:

1. Obtain a one-foot length of two-wire insulated electric cable, similar to that used for table lamps or small extension cords. Cut straight across one end of the wire, leaving no bare metal wire extending past the rubber insulation on either of the two wires. At the other end of the cord, strip a short bit of insulation from the two wires. Attach the first of the two bare-wire leads to a terminal on the igniter. Attach the second bare lead to the other terminal. Secure the wires with electrical tape.
2. Test the operation of the igniter. When the button is pressed, a very tiny spark should jump across the square-cut end of the cable.
3. Use a nail or awl to punch a hole in the plastic lid of the film canister. Insert the end of the cable through the hole. Adjust the cable so it extends to the center of the attached can.
4. Place only two or three drops of methyl alcohol in the canister and attach the lid. Shake the canister for a few seconds.
5. Press the igniter button. A small pop is emitted as the canister snaps away from the lid.
6. Do not use more than a few drops of alcohol. Excess alcohol may ignite and drip from the can. Practice this activity thoroughly.

The Science:

Alcohol is a member of a family of chemicals called hydrocarbons. Most compounds commonly used for heating, as solvents or for transportation fuels are hydrocarbons. Natural gas (methane), gasoline, turpentine, ether, kerosene, paraffin, diesel fuel, propane, butane and alcohol are a few examples. Made mostly of hydrogen and carbon atoms, almost all hydrocarbons are combustible. In most cases the products of hydrocarbon combustion are carbon dioxide and water vapor.

The one or two drops of alcohol placed in the plastic container quickly evaporate. As alcohol vapor mixes with oxygen in the air, an explosive mixture is created. Although some of the oxygen in the container is consumed in the reaction, carbon dioxide gas and water vapor are vigorously released, with the resulting pop!

Note:
-When someone inevitably requests to add more alcohol, ask them to predict the change in combustion rate. More alcohol does not create a louder pop. Only the alcohol that evaporates is involved in the explosion. Excess alcohol merely floods the canister.

Sketches & Observations:

Notes:

INTERNAL REFLECTION

What is it? A glass of water and a pencil are used to demonstrate how light travels through optic cables

-Try different clear liquids and compare the outcome. (alcohol, corn syrup, paint thinner— be careful)

What you'll need:

Checklist:

- ❑ A transparent drinking glass
- ❑ Water
- ❑ Pencil or other slender stick

Challenge:
-Make an observation.
Where is light being bounced (reflected)?
Where is it being bent (refracted)?
Where can you position the pencil in the water to get the most refraction or, the least?

Hint:
-Use a plain glass with no patterns or printing.

Here's How:

1. Fill the glass three-quarters full with clean water. Hold the glass so the underside of the water surface can be observed through the side walls of the glass.
2. Slowly lower a pencil into the water. Make note of the nature of the of reflected pencil image appearing on the underside of the water.

The Science:

The boundary between the water surface and air acts as a mirrored surface due to the fact that water has a higher index of refraction, or tendency to bend the path of light. If light should strike the water surface above a certain threshold angle, called the critical angle, it will pass through the water surface. Below the critical angle, it is reflected as if it were bouncing off a mirrored surface.

Sketches & Observations:

Notes:

Gravity Strip

What is it? Using gravity to test your reflexes!

Turn off the fan.
Blowing wind = poor results.

What you'll need:

Checklist:

- ❑ Stiff paper or cardboard
- ❑ Marker pen
- ❑ Ruler
- ❑ Scissors

Something to ponder:

- Why are the marks not equally spaced?

- Practice is required.

Here's How:

1. Cut strips of cardboard to these dimensions: 2.5 cm x 30 cm.
2. From one end, measure and make corresponding time markings at the following distances: at 0.0 cm mark 0.0 sec, at 5.0 cm mark 0.06 sec, at 10.0 cm mark 0.12 sec, at 15.0 cm mark 0.17 sec, at 20.0 cm mark 0.21 sec, at 25.0 cm mark 0.24 sec, at 30.0 cm mark 0.26 sec.
3. This activity requires at least two people; a dropper and a catcher.
4. The dropper should hold the strip vertically between two fingers, with the zero marks at the lower end of the strip. The catcher should hold two fingers apart over the zero marks, or starting position.
5. When the dropper releases the paper strip, the catcher should pinch it as fast as possible; estimating their catch time if they should capture it between markings.

The Science:

The constant pull of the gravity causes objects to fall towards the earth's center at a rate of 9.8 meters per second per second; or 9.8 m/sec^2. Simply stated: The farther an object falls towards the earth, the faster it goes. It accelerates.

Every object in the universe exerts gravity; the earth.... the moon.... you! The earth's gravity pulls on the moon, keeping it in orbit. The moon's gravitational pull on the earth actually lifts ocean water a bit, creating tides and waves.

Sketches & Observations:

Notes:

Hot Hands

Are you hot-blooded? Build a simple thermometer to find out.

What you'll need:

Checklist:

- ❑ A one-liter or larger flask or glass bottle
- ❑ One-hole stopper to fit
- ❑ Length of glass capillary tubing or other transparent tubing with small inside diameter
- ❑ Food coloring
- ❑ Water

Optional:

- ❑ Support stand

Notes and hints:
- Select a bottle that fits your hands.
- The bottle walls must be stiff; a flexible bottle will not work.
- No leaks! Check for air leaks.

Here's How:

1. Make a hole in the bottle stopper sized to fit the tubing. Insert the tubing and position it just above the bottom of the closed bottle.
2. Use glue or caulk to make an air-tight seal between the tubing and the stopper.
3. Fill the bottle one-third full with heavily colored water. Insert the stopper and tubing into the mouth of the bottle.
4. Carefully blow through the tube.... just a little! You need to "charge" the space above the water with some extra air. If you blow in too much air, you'll get sprayed with colored water. Blow in just enough to lift the water midway up the tube.
5. Check and make certain that, at room temperature, the screw head will just barely pass through the screw eye.
6. Wrap your hands around the bottle to transfer body heat to the trapped air within it. Observe any changes in the water level.
7. Option: Tape a card strip to the tube and make marks to denote temperature changes.

The Science:

Adding heat energy from any source causes an increase in molecular motion of the air trapped in the flask. The resulting increase in air pressure on the surface of the colored water causes it to rise in the tubing. Capillary (small-diameter hole) tubing is suggested because its relatively small inside diameter amplifies the change in water level.

Something to ponder:

From where does the heat in your hands come?

Sketches & Observations:

Notes:

Tasty Candles - A Neat Science Trick

What is it? Make candles that you can eat! Tasty!

What you'll need:

Checklist:

- ❑ A large baking potato
- ❑ Almond or Brazil nut slivers
- ❑ Cork borer or apple corer
- ❑ Matches

Hints and notes:
-Keep your audience at a distance.
-Present this in a dimly lit room.
-Pull the fake candle from a pile of real candles.

Here's How:

1. Caution: For this science trick to be successfully presented, you should practice it several times.
2. A set of metal tubes of assorted diameters, cork borers are used in laboratories to cut holes through corks and other stoppers. Borrow a three-quarter-inch or one-inch-diameter cork borer from a chemistry teacher. Hardware stores often have apple coring tubes. Those work just as well.
3. Pushing through the long axis of a potato, use the borer to cut a cylinder of potato. Make a straight cut at both ends of the potato cylinder. From a distance, it resembles a wax candle.
4. Create a fake candle wick by inserting a sliver of almond or Brazil nut at one end of the potato cylinder. Hold the fake candle upright and ignite the nut.
5. From a distance, the potato and burning sliver nut give the appearance of a real candle.

6. Blow out the flame on the nut wick. Wait a few seconds for the burnt nut to cool and bite off the end of the potato. Chewing loudly enhances the dramatic effect.
7. Presentation: Light the fake candle. Suggest to your audience that humans need to eat food as a source of energy.
8. Blow out the fake candle. While informing the class that wax is one of your favorite energy sources as well as a great snack, bite off a chunk and chew it loudly.
9. Inform your audience of the composition of your fake candle, reminding them of the hazards involved with repeating the activity.

The Science:

The secret is in the "wick." Most nuts contain a significant amount of oil. Like most oils, it will burn in air, releasing its stored energy in the form of heat and light. Similarly, we eat nuts not only for their good taste, which comes from the oils, but for the energy that oil provides our bodies.

Sketches & Observations:

Notes:

Two-Temperature Water

What is it? A simple and wet science activity designed to make you ponder.

What you'll need:

Checklist:

- ❑ Bowl of ice water
- ❑ Bowl of very warm water
- ❑ Bowl of room-temperature water

Hints and notes:

-Keep a towel handy.
-Hot water.... but, not too hot.

Here's How:

1. The bowl of warm water should be as warm as possible without being too hot to receive a person's hands.
2. Place a "cold" sign on the bowl containing ice water. Place a "hot" sign on the bowl containing warm water. Place a "?" sign on the bowl containing room temperature water. Do not divulge to your audience the temperature of the contents of any of the bowls.
3. Ask a volunteer to feel the contents of the "cold" bowl and report the results .
4. Ask the volunteer to feel the contents of the "hot" bowl and report the results.
5. Next, ask the volunteer to soak one hand in the cold water and the other hand in the warm water.

6. While the volunteer's hands are soaking in the water, explain what is to happen next. Direct the volunteer to remove both hands at a given signal from the two bowls and plunge them simultaneously into the "?" bowl. Explain to everyone that the "?" bowl contains room-temperature water.
7. Ask anyone watching to predict what the volunteer will feel when he or she plunges both hands into the "?" bowl. Give them time to discuss the possibilities and ponder them a bit.
8. Ask the volunteer to describe what he or she is feeling. What is the relative temperature of the "?" bowl? (To the warmed hand, the water feels cold. To the chilled hand, the water feels warm.)

The Science:

Human skin contains nerve receptors capable of detecting a variety of stimuli: hot, cold, pressure, pain. Receptors that are detectors of hot and cold sensations are sensitive to the flow of heat energy.

Scientists experienced with thermodynamics understand that heat always flows from areas of higher temperatures to regions of lower temperatures. Matter never "gets cold." It either gains or loses heat energy.

The chilled hand receives heat energy from the room-temperature water, thereby detecting a flow of heat energy into the hand and sensing the water as warm. The warmed hand gives heat energy to the room-temperature water, thereby detecting a flow of heat out of the hand and sensing the water as cool.

Sketches & Observations:

Notes:

Vacuum-Packed Friends

What is it? A demonstration of the massive strength of the air around us.

Hints and notes:
- Never do this activity alone.
- Keep the plastic bag away from your face at all times.
- Use a thin, low mil bag.

What you'll need:

Checklist:

- ❑ Large plastic trash bag
- ❑ Tape
- ❑ Vacuum cleaner

- Allow your volunteer to remain in the bag for less than a minute or so. If he feels even slightly uncomfortable, discontinue the activity.

Here's How:

1. Remember: Plastic bags can be dangerous for small children.
2. Direct a volunteer to step into a plastic trash bag and squat on the floor. Carefully gather the mouth of the bag around the volunteer's neck and seal with tape. Make certain the volunteer is comfortable and has no problem breathing. The bag should not extend near the volunteer's face.
3. Tear a small hole in the side of the plastic trash bag and insert the hose of a vacuum cleaner. Use tape to seal the bag to the pipe.
4. When the vacuum cleaner is switched on, a partial vacuum is created in the plastic bag. The bag will collapse against the volunteer, making him or her appear to be shrink-wrapped and very immobile.

The Science:

We live at the bottom of an ocean of air. The weight of the air above us exerts almost 15 pounds of pressure per square inch. Under normal conditions, the total pressure exerted on the surface area of the midsized human body exceeds 50,000 pounds.

The vacuum cleaner does not create a perfect vacuum in the bag. The partial vacuum created is sufficient to dramatically demonstrate the pressure exerted by ambient air. It is the air on the outside of the bag pushing against the volunteer that prevents him from moving his arms and legs.

Sketches & Observations:

Notes:

FLOATING PENNIES

What is it? The zinc core of a penny is chemically removed leaving a hollow copper shell.

What you'll need:

Checklist:

- ❑ Tall glass cylinder
- ❑ New penny
- ❑ Small metal file
- ❑ Diluted hydrochloric acid
- ❑ Baking soda

Something to ponder:
-Do you observe anything that might suggest a chemical reaction is taking place?

Here's How:

1. A very weak concentration of acid is required for this activity. However, safety protocol should be followed at all times. Make sure proper eye protection is provided.
2. You may be able to locate a 5% solution of hydrochloric acid at a pharmacy or science supplier. A 0.1m concentration will work just as well. Muriatic acid, a form of hydrochloric acid used to etch and clean concrete, is usually sold in paint and hardware stores. Most muriatic acid is sold at 10% concentration. For this activity, 10% muriatic acid should be diluted fifty-fifty with distilled water.
3. Use a triangular or other sharp file to cut several notches into the edge of a new penny. Cut the notch deep enough to expose the zinc layer under the copper cladding. Pour some acid into a clean glass container. Drop two or three notched pennies into the acid and observe. Leave the pennies in the acid overnight. Do not seal the container. Allow it to "breathe."
4. Use tongs or another safe method to remove the pennies. Wash the pennies thoroughly before inspecting them.

The Science:

One of the defining characteristics of the group of chemical elements called metals is their tendency to react with acids with a corresponding release of hydrogen. You may notice hydrogen gas escaping from the interior of the penny coin. Hydrogen bubbles may even cause the pennies to float in the acid.

Chemists coined the phrase "electromotive series" to describe the tendency of metals to lose electrons in a chemical reaction. Zinc is listed higher than copper on a chart of the electromotive series. It has a greater tendency to lose electrons than copper, thus being more reactive with the acid.

Sketches & Observations:

Notes:

Microwave Light Bulbs

What is it? Light up a variety of light bulbs without applying normal electric current.

What you'll need:

Checklist:

- ❑ Circular or short fluorescent bulb
- ❑ Assorted incandescent light bulbs
- ❑ Microwave oven
- ❑ Cup of cool water

Something to ponder:
-Do some regions of the microwave oven cavity work better than others?

Here's How:

1. Microwave ovens should not be operated dry. Most require a minimum load equivalent to 50 ml of water. Place a cup of water in the microwave oven to fulfill that requirement.
2. Obtain a fluorescent bulb that easily fits within the operating cavity of the microwave oven. Circular bulbs work well. Place it in the oven and turn the oven on for a few seconds at a time. If there is any sparking or other atypical operation of the oven, discontinue this activity immediately.
3. In the presence of sufficient microwave radiation, the bulb should emit bright light. Often the bulb emits more light than during normal operation.
4. Repeat with a variety of incandescent bulbs or multiples of same.

The Science:

A device above the cavity in microwave ovens, commonly called a magnetron, generates electromagnetic radiation. The radiation generated easily passes through food items. Most foods contain water. The wavelength of microwave radiation corresponds to the natural vibration of water molecules. The applied energy causes water molecules in food to increase their vibrational and rotational motion. They "heat up."

Radiation generated by the magnetron also excites phosphors in the light bulb, causing light to be emitted. That same radiation can cause electrons in a filament to move back and forth through the filament wire so fast that the filament glows.

Sketches & Observations:

Notes:

Puffed-Rice Fleas

What is it? Kernels of puffed rice cereal behave like fleas when charged with static electricity.

What you'll need:

Checklist:

- ❑ Glass or porcelain dinner plate
- ❑ Puffed rice cereal or small styrofoam bits
- ❑ Balloons or a length of PVC pipe
- ❑ Plastic comb
- ❑ Charging cloths (wool, plastic wrap, dry cotton rag)

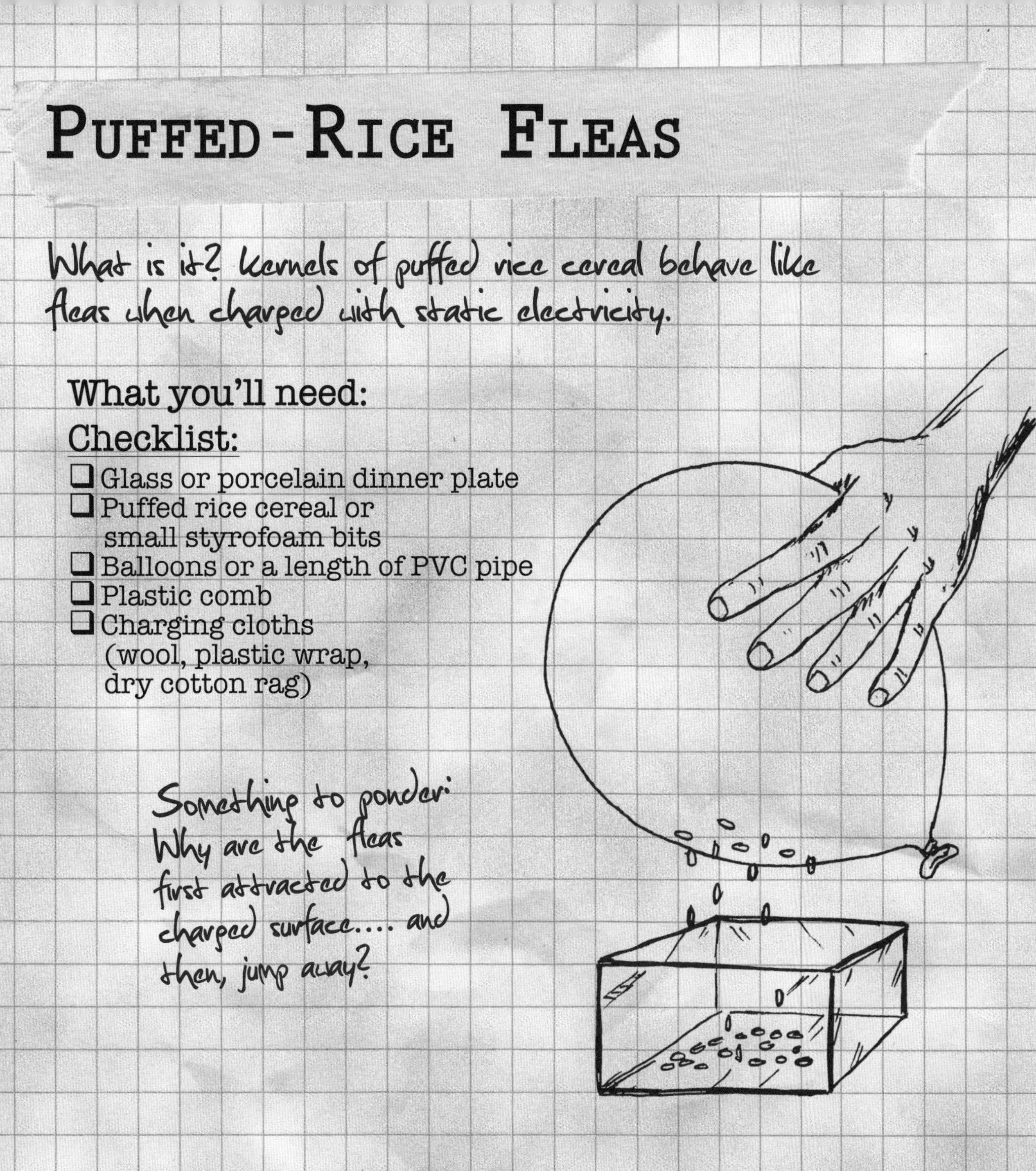

Something to ponder: Why are the fleas first attracted to the charged surface.... and then, jump away?

Here's How:

1. Sprinkle several kernels of puffed rice cereal on the dinner plate. Use a charging cloth to stroke an inflated balloon in one direction.
2. Bring the charged balloon near to the puffed rice. The kernels of cereal will jump off of the plate and adhere to the surface of the balloon.
3. Hold the balloon motionless. Some of the kernels of puffed rice will jump away from the balloon.
4. With practice, the balloon can be rotated immediately after collecting several kernels of rice, positioning the rice on the upper surface of the balloon. This should dispel the notion the rice is merely falling off the balloon rather than being repelled from it.

The Science:

Puffed rice is a poor conductor. Similar to a charged balloon, an excess of electrons will collect on the surface of the kernel when it is charged with static electricity.

When a charged balloon is initially brought near the kernels of puffed rice, the "unlike charged attract" rule is demonstrated. The balloon is negatively charged with an excess of electrons. The rice has no charge. However, compared to the balloon, the rice has a relative positive charge. There exists sufficient difference in charge to cause the rice to be attracted to the balloon.

Eventually some of the excess static charge on the surface of the balloon migrates to the rice, causing the rice to become negatively charged. The "like charges repel" rule is then demonstrated. Kernels of rice jump from the balloon.

The "fleas" are jumping!

Sketches & Observations:

Notes:

Gremlins in a Can

What is it? A great science trick.... gremlins living in a paint can.

What you'll need:

Checklist:

- ❑ A one-quart empty paint can, with lid
- ❑ Styrofoam peanuts and cups
- ❑ Tweezers or tongs
- ❑ Acetone
- ❑ Awl, or hammer and nail

Something to ponder:
Make up a descriptive story of the imaginary creatures living in the can.... with voracious appetites for styrofoam.

GREMLIN FOOD

GREMLIN

Here's How:

1. Use an awl or a hammer and nail to punch "breathing holes" in the lid of an empty quart paint can. You may decorate the exterior of the can to give the impression you have living organisms in the can.
2. Pour acetone into the empty can, about 1 cm deep. Remember to keep flames away from the mouth of the can. Acetone is flammable.
3. As a test, drop a few styrofoam peanuts or bits of a foam cup into the acetone. Notice that it bubbles and hisses and appears to completely dissolve in the clear acetone.
4. Amazingly, you will be able to deposit several gallons of peanuts into the one-quart can! Notice the styrofoam turns to a thick putty at the bottom of the acetone. Use tongs to remove the goo. Place it on a piece of kitchen foil. Allow the acetone on it to evaporate. The putty will quickly harden into a hard plastic disc.

The Science:

Styrofoam is mostly air or other gas.... about 98% gas trapped in zillions of tiny bubbles. The skin of the tiny bubbles is made of a plastic called polystyrene. Acetone breaks some of the chemical bonds holding polystyrene molecules together. In that process the trapped gas is released, leaving a mass of polystyrene goo. Those messy gremlins!

Sketches & Observations:

Notes:

Fire Tornado

What is it? A tall flame vortex is created in a spinning wire cage.

Something to ponder:
What is the role of the screen cage in the creation of the flame vortex?

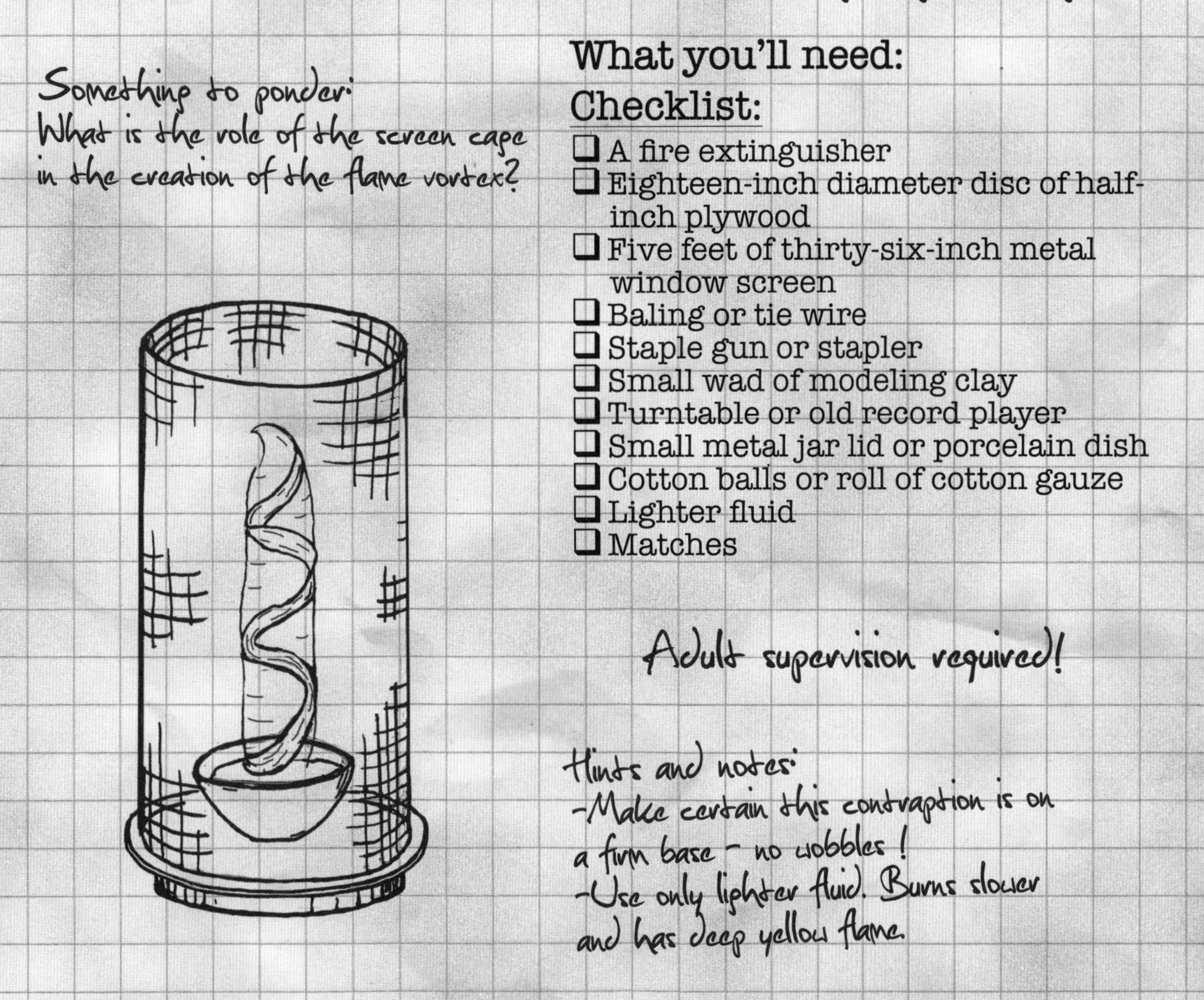

What you'll need:

Checklist:

- ❑ A fire extinguisher
- ❑ Eighteen-inch diameter disc of half-inch plywood
- ❑ Five feet of thirty-six-inch metal window screen
- ❑ Baling or tie wire
- ❑ Staple gun or stapler
- ❑ Small wad of modeling clay
- ❑ Turntable or old record player
- ❑ Small metal jar lid or porcelain dish
- ❑ Cotton balls or roll of cotton gauze
- ❑ Lighter fluid
- ❑ Matches

Adult supervision required!

Hints and notes:
- Make certain this contraption is on a firm base – no wobbles!
- Use only lighter fluid. Burns slower and has deep yellow flame.

Here's How:

1. Locate the exact center of the plywood disc. Drill a small hole at the center so the disc will rest flat on the turntable. Use a staple gun or stapler to attach the wire screen to the disc, forming a cylindrical tube. While maintaining the cylindrical shape, use wire or staples to stitch together the loose flaps of screen.
2. With a fire extinguisher handy, place the cage on the turntable and ignite the wick. Turn on the turntable to 33 rpm. In a few seconds the fluid flame will create a flame vortex resembling a fire tornado!
3. Be careful! Use only the prescribed quantity and type of fuel.

The Science:

Hot gases from the combustion products tend to rise. They are less dense and pushed upward by more dense gases in the cooler surrounding atmosphere. Acting as a centrifuge, the rotating screen causes rotation of the gases within the cage. The unheated air around the flame is denser and by the action of centrifugal force is "pushed" outwards more than the less dense gases of the flame. This outward flow of air creates a pocket of lower pressure above the flame, allowing it to rise higher than normal. In this context, "rise" is not completely accurate. Less dense warm air is pushed up by nearby dense cold air. It does not rise on its own accord.

The flame vortex created by this activity has some similarity to naturally occurring tornadoes. Under certain conditions, warm and moist atmospheric gases can be temporarily trapped under a fast moving front of cold air. When a "hole is poked" through the covering blanket of cold air, warm and less dense trapped air rises rapidly through that hole. Sliding action of the moving front causes a rotating action and a vortex is created.

Sketches & Observations:

Notes:

Salt Engineering

What is it? A simple cardboard tube and some table salt are used to demonstrate a basic concept in engineering. An amazing trick!

What you'll need:

Checklist:

- ❑ Cardboard tube, similar to a wax paper tube
- ❑ Tissue or wax paper
- ❑ Rubber bands or tape
- ❑ Table salt
- ❑ One-inch dowel or broomstick

- Be prepared to clean up spilled salt.... just in case!

Something to ponder:
-Shouldn't the extra weight of the added salt make it easier to tear the paper? What gives?

Note:
-Look for engineering structures designed to redirect a force or load.

Here's How:

1. Cut tissue into six-inch squares. Use rubber bands or tape to attach the paper over one end of the cardboard tube, creating a paper membrane or drum head.
2. For a control test, push or drop a wood dowel through the tube and burst through the paper cover. Observe how easily the dowel pushes through the thin paper.
3. Remove the damaged tissue and replace it with another piece. Pour salt into the tube to a depth of at least three inches.
4. Attempt to burst the paper by pushing the dowel on the surface of the salt. Is it easier or more difficult to break the paper?

The Science:

The tiny and numerous salt crystals transfer most of the applied force to the walls of the cardboard cylinder. You might be able to feel the pressure of the salt pushing out as you push the dowel with your other hand.

Trusses and beams designed for large building structures employ similar effects in order to spread or change the direction of forces and loads.

Sketches & Observations:

Notes:

Bubble Belly Dancers

What is it? Unusual wave patterns are created with a soap film.

What you'll need:

Checklist:

- ❑ Distilled water or very clean tap water
- ❑ Dish detergent; unscented Dawn works well
- ❑ Glycerin
- ❑ Saucers or shallow bowl
- ❑ Metal jar lids
- ❑ Tape
- ❑ Dropper pipettes
- ❑ Spoons
- ❑ Mixing cups

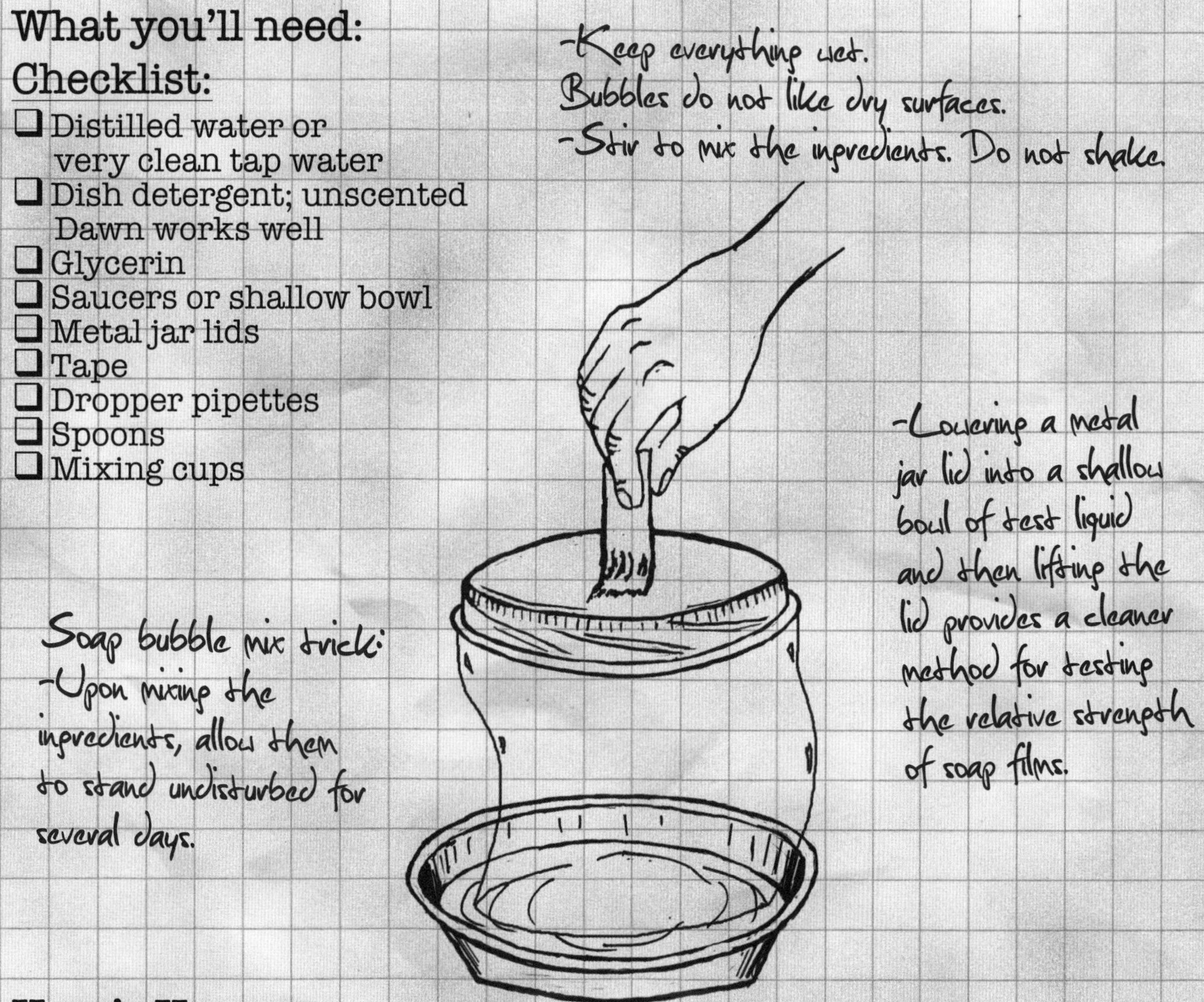

Here's How:

1. A good mixture for making soap films and blowing bubbles can be made thus:
   Clean water: 100 parts
   Dawn liquid: 10 parts
   Glycerin: 2 parts
2. Fill a shallow saucer with aged bubble mix. Lower the open mouth of a jar lid into the solution then slowly lift it out of the mix. You should be able to create a bubble cylinder about three or four inches tall. Slight vibrations in your hand will cause the cylinder move in a wild wave motion.... a bubble belly dancer.

The Science:

Detergents are designed by chemists to be attracted to water molecules. That attraction creates the skin or film we call bubbles.

Tap water and mineral water both contain compounds that interfere with the intermolecular attraction that creates surface tension. Contaminants in the water decrease surface tension.

Glycerin is a long organic molecule slightly soluble in both detergent and water. A minute amount dissolved in the soapy solution increases the strength of soap films.

Sketches & Observations:

Notes:

A Haunted Bathroom

What is it? A steamy bathroom reveals the handwriting of ghosts that visited while you were in the shower!

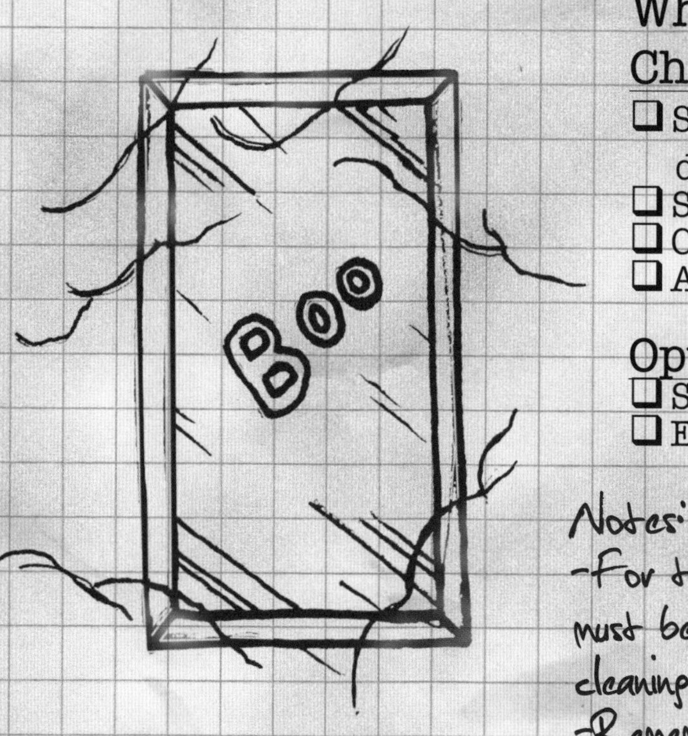

What you'll need:

Checklist:

- ❑ Small container of clean or distilled water
- ❑ Soap or detergent
- ❑ Cotton swab
- ❑ Alcohol

Optional:

- ❑ Stick or dowel rod
- ❑ Electric hair dryer

Notes:
-For this to work, the bathroom mirror must be spotlessly clean. No soap or cleaning agent residue, either.
-Remember.... little is better. Use only the slightest bit of soap solution.

Here's How:

1. This is a great activity to perform at home. However, it does require some practice to develop a successful soap writing technique. Make several test trials before haunting your family and friends.
2. Make certain the mirror is clean. Use alcohol on a paper towel to remove any soap film.
3. Dissolve one or two drops of detergent or a tiny bit of soap in some clean water. Moisten the end of a cotton swab with the solution.
4. Using the swab, write a secret message on a mirror or other large glass surface. Allow the message to dry. Use a hair dryer to speed the drying process.
5. Once the water has evaporated, a very thin layer of soap remains on the surface of the glass. If too much soap was used in the solution the dry message will be detectable. If that is the case, wash the glass and repeat with a less concentrated solution.

6. Likewise, if too little soap is used, the message will not appear in the steamy bathroom. Test your solution before attempting to confound your message recipient.
7. Allowing plenty of time for it to dry, the message should be written on the bathroom mirror prior to the family's routine schedule for bathing and showers. When a family member enters the bathroom the mirror should appear normal. After a steamy bath or shower, they will step in front of the mirror to find a message written to them.
8. There might be a cry through the house: "Who sneaked into the bathroom and wrote this message while I was in the shower?"

The Science:

The polar nature of water molecules causes them to be attracted to each other. In a steamy bathroom, water vapor suspended in air collects to form small droplets. Adhesive attraction between water and glass causes those droplets to cling to the surface of the mirror, creating the characteristic fogged-over appearance.

Soaps and detergents interfere with the intermolecular attraction between water molecules. Water droplets deposited on regions of the mirror surface previously coated with soap solution collapse and spread over the surface of the glass. This action creates the illusion that someone has used their finger to write a message on the fogged-over mirror. Yikes!

Sketches & Observations:

Notes:

Gelatin-Optic Fibers

What is it? Strips of gelatin dessert and a laser pointer demonstrate total internal reflection.

What you'll need:

Checklist:

- ❑ An inexpensive laser pointer
- ❑ Package of unflavored gelatin or light-colored gelatin dessert
- ❑ Shallow cake pan
- ❑ Spatula
- ❑ Mixing bowl, water, etc.
- ❑ Protractor

Hint:
-This works best in a darkened room.

Here's How:

1. Follow the package directions for mixing the gelatin, using only half the suggested amount of water. The resulting thicker-than-normal gelatin is more suitable for this activity. Pour the gelatin mixture one inch deep in a shallow pan and refrigerate.
2. Cut the stiffened gelatin into one-inch-wide strips, creating strips as long as possible. Use a spatula to place a gelatin cable on a flat surface or table top.
3. Dim the room lights. Point the laser beam through one end of the gelatin strip. Position the laser so the beam will reflect several times off of the interior side walls of the gelatin.
4. Use a protractor to measure the angles of incidence and reflection. The angles should be equal.

The Science:

Extremely pure glass used in the manufacture of fiber-optic cable allows light to pass through undisturbed for long distances. Light traveling through the glass stays inside because it reflects off of the internal surface of the glass.

The boundary surface between the gelatin walls and air also acts like a mirrored surface due to the fact the gelatin has a higher index of refraction, or tendency to bend the path of light. If light should strike the side walls above a certain threshold angle, called the critical angle, it will pass through the side walls. Below the critical angle, it is reflected as if it were bouncing off a mirrored surface. The critical angle for internal reflection in the gelatin can be observed by adjusting the position of the laser.

Sketches & Observations:

Notes:

Sandwich-Bag Dartboard

What is it? An amazing science trick: Sharp pencils and water filled sandwich bags demonstrate the nature of certain plastic polymers.

What you'll need:

Checklist:

- ❑ Pint or quart zip-lock plastic bags
- ❑ Sharpened pencils
- ❑ Water
- ❑ Sink or spill pan

Notes:
-Practice this activity over the sink or a pan.
-Be careful. Don't stab yourself.

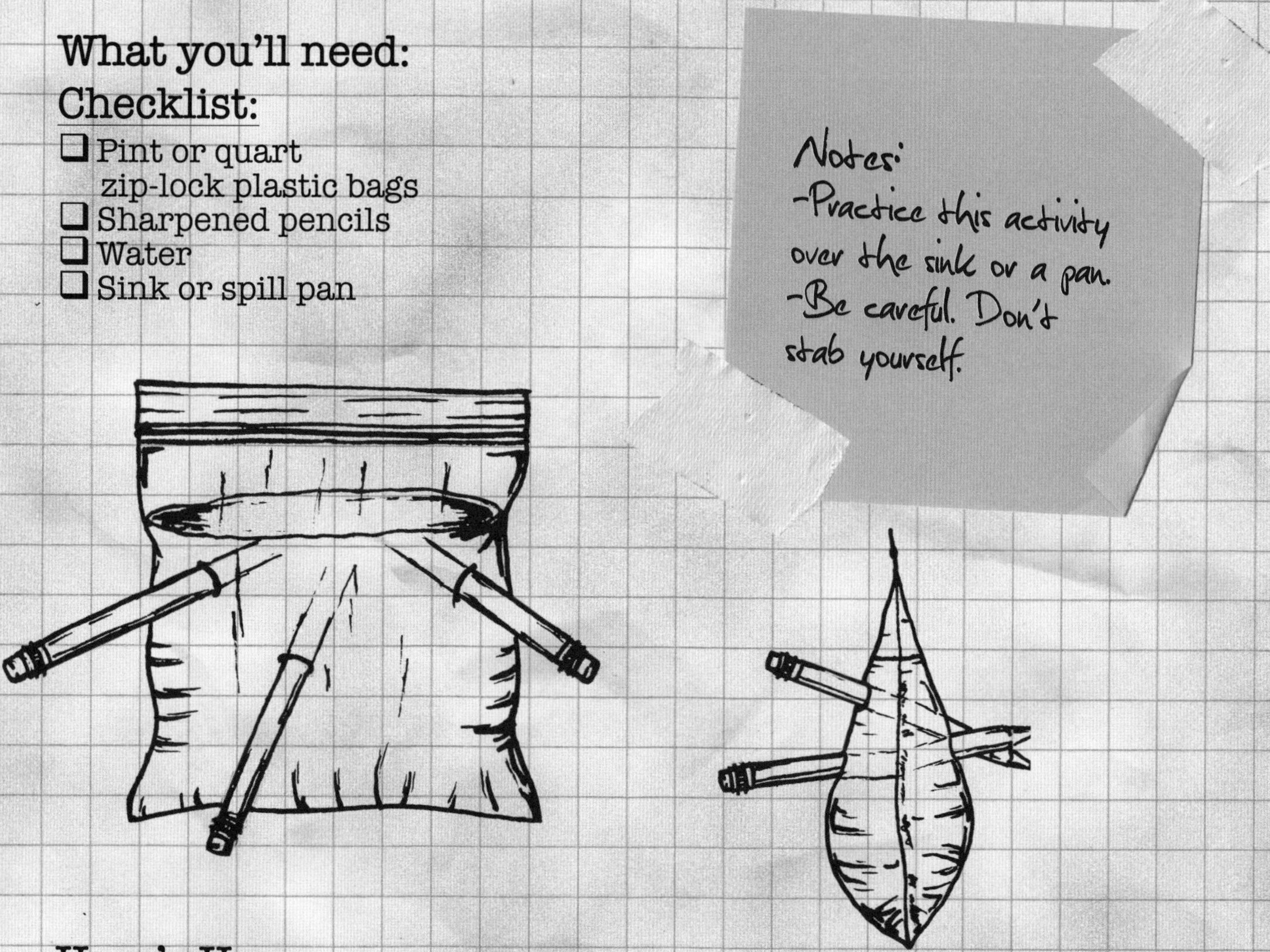

Here's How:

1. During the activity, you will be pushing sharpened pencils through water-filled plastics bags. Make provisions for spilled water. Follow proper safety protocol.
2. Fill a bag three-quarters full with tap water and seal it shut. While holding the bag by the sealed mouth, and allowing it to hang vertically, quickly thrust a sharpened pencil through both walls of the bag below the water line.
3. Do not withdraw the pencil. With practice, you should be able to perform the feat with the loss of only a few drops of water.
4. Repeat the action with several pencils.

The Science:

Most materials commonly known as plastics are made of very long molecules called polymers. The material used to make sandwich bags, polyethylene, consists of polymer molecules intertwined and linked to form a web-like matrix. That molecular web gives polyethylene its characteristic resistance to tearing. Polymers with different molecular structures exhibit different physical and chemical characteristics.

Sketches & Observations:

Notes:

Tree Drinking

What is it? Learn a survival technique:
Gather pure drinking water from a tree.

What you'll need:

Checklist:

- ❑ Thin plastic painter's tarp or a very large plastic trash bag
- ❑ Small rubber bands
- ❑ Tape or tie strips
- ❑ A tree or large shrub

Hint:
-Leave the bag on the tree branch no more than four or five hours.
-Expect to gather between a cup and a quart of water.

Something to ponder:
-Do you think this activity will work better in sunlight, or at night?

Here's How:

1. Wrap the plastic sheet or bag around the end of a large leafy branch. Use tape or tie-wire to seal the plastic tight around the branch stem, creating an airtight pouch.
2. Wait.
3. Make frequent observations of the interior surface of the plastic. You should notice water droplets condensing on the interior surface. Eventfully they will run down and pool in a corner of the plastic.

The Science:

In their growing process, trees and shrubs draw a large amount of water from the earth into their roots and up to their leaves. In a process called transpiration, pure water vapor is emitted though small openings (stomata) on the underside of leaves. That water vapor collects on the plastic and you can capture it.

Sketches & Observations:

Notes:

Chain-Loop Ecology Hunt

What is it? A miniature safari.... in your own backyard.

What you'll need:

Checklist:

- ❑ A three-foot length of chain
- ❑ Magnifying glass
- ❑ Tweezers/forceps
- ❑ Notepaper and pencil
- ❑ Piece of wool cloth, glove or sock
- ❑ Large spoon or hand trowel

A blade of grass, part of a leaf, a grasshopper leg, a worm casing, some ants, a seed of some sort, a bit of spider web an aphid....

-Ask for help if you cannot identify a specimen.
-Work slowly and carefully. Take clear and accurate notes.

Something to ponder:
-Are there areas of your yard containing a noticeably different number of species? Why?

Here's How:

1. Use a bit of wire or paper clip to attach the two ends of the chain, forming a loop. Locate an interesting area in your backyard and toss the chain loop on the ground. Form into a circle.
2. From a comfortable position carefully inspect the interior of the circle. Locate specimens of each plant and animal species living within the bounds of the circle, now and in the past. Place your evidence in an orderly fashion on the cookie sheet. You may need to use the magnifying glass to find evidence of life forms.
3. If you cannot find an intact specimen, look for evidence that it was once in the circle.
4. Make a list of the life form evidence you locate in the region.

The Science:

There have been identified approximately one- to one-and-one-half-million species of animals on the planet Earth, and 950,000 of those are insect species! In addition, hundreds of thousands of plant species have been identified. It is estimated that many millions of species live upon the earth. That is why you might find evidence of a large number of species living within your small backyard circle. The last time I did this with my grandkids, we found 22!

Sketches & Observations:

Notes:

Impossible Rope Trick

What is it? A classic problem-solving trick that will keep your friends tied up in knots.

What you'll need:

Checklist:

- ❑ Two four-foot lengths of rope
- ❑ Two volunteers

Hints and notes:
- Use rope that is about one-quarter-inch in diameter.
- Smooth cotton rope works best.
- Similar to window sash cord.
- String is difficult to control and makes it difficult to see what you are doing.

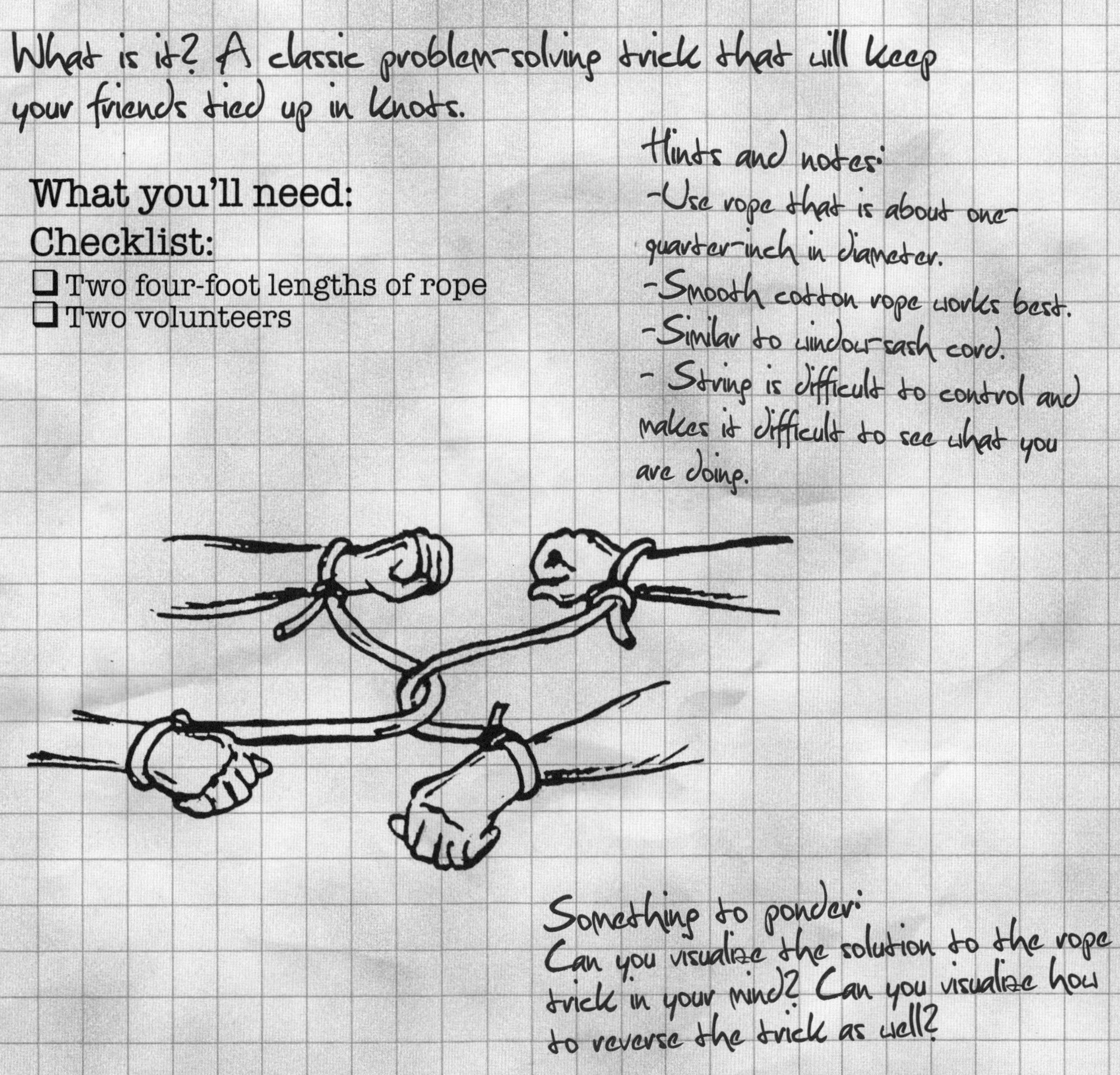

Something to ponder:
Can you visualize the solution to the rope trick in your mind? Can you visualize how to reverse the trick as well?

Here's How:

1. Tie an end of a rope around the wrists of your volunteers as shown. Do not attach the rope tightly. Leave the tied loops fairly loose.
2. Challenge your friends to separate themselves.... without removing the ropes or untying the knots holding them to their wrists. Ask them to mentally picture the solution in their head.
3. Solution: Direct one volunteer to grab the midpoint of her rope and slide it under the loop tied to her friend's wrist. Next, the volunteer should lift that midpoint over her friend's hand and pull it back under the loop. They will be separated.

The Science:

The purpose of this activity is twofold: to practice problem solving, and to practice visualizing solutions in your mind. Those are two of the many thinking skills used by scientists around the word. The highly regarded scientist Albert Einstein often referred to his "experiments of the mind." He was a master at visualizing problems in his head.

Sketches & Observations:

Notes:

Thinking Like a Scientist

Scientists in all fields of study use the same set of skills to perform their work. You will find that chemists, biologists, physicists, astronomers and geologists are experts at "science process skills."

Each of the activities in this book includes at least one of those skills. As you perform an activity, you will be practicing the use of that skill, making you a better scientist.

Here is a short list of some science process skills and suggestions on how to practice them:

Data recording and analysis: Keep good and timely notes while you work on a science project. Make note of both successes and failures. With practice, you will be able to ponder your notes and perhaps find the causes of problems, or find better methods of performing the activity.

Measurement: Practice using all sorts of devices to take measurements: rulers, yardsticks, thermometers, scales and balances, etc. Don't just guess. Measure it!

Observation: Use your eyes and your brain to thoughtfully gather information about an object or event. Scientists do not merely "look" at an object; they observe it, making mental notes of what they observe.

Prediction: Scientists rarely just guess. More often, they predict. A prediction is a guess based on past experience. Scientists stop and think before they speak or make a guess.

Experimenting and testing: Scientists do not throw things together to see what happens. They experiment by testing one thing at a time, while they keep everything else constant, or the same.

Calculating: All scientists are capable of performing a variety of mathematical functions. It's OK to use a calculator, but you should practice calculating on your own from time to time.

Manipulating materials and equipment: Most scientists are "hands-on" types of folks. They constantly get trained or train themselves to use all sorts of tools, materials and equipment. Practice. Practice. Practice. Doing so will allow you to be a master science equipment operator.

Adapting: Scientists have the ability to change to the situation, while staying focused on an original objective. They respond well to, "Oooops, that wasn't supposed to happen." And scientists will often say, "Let's try it again."

Precision and accuracy: Scientists are careful. They tend to details and specifics in all aspects of an activity. They watch what they are doing.

Scientist Tool Kit Ideas for a Young Scientist

Here's a list of items you might want to collect and keep in a tool box or bag. They'll be handy for all sorts of scientific investigations.

1. Magnifying device, hand lens 10X or less
2. Forceps, tweezers
3. Probe; a pointed stick
4. Spatula
5. Safety glasses
6. Collection containers; pill bottles, zip-lock bags
7. Pen and notepad
8. Tape measure, ruler
9. Gloves
10. Pipette, dropper
11. Stopwatch / timer
12. Thermometer
13. Magnet
14. Clamps
15. Wiping cloth
16. Tape; masking, electrical, duct
17. Marker, permanent
18. Labels
19. Candle
20. Flashlight; optional red lens for night use
21. Magnetic compass

How Hot Is It?

Scientists are experts at estimating. In order to make an accurate estimate, scientists look for known reference points to which they can make reasonable comparisons.

Here are some reference points for common temperatures. Referring to them will help you make better heat-related estimates.

Absolute zero	-273 °C / -460 °F
Liquid nitrogen	-210 to -196 °C / -346 to -321 °F
Dry ice	-78.5 °C / -109 °F
Ice water	0 °C / 32 °F
Your body (internal)	37 °C / 98.6 °F
Hot water from a sink tap	60 °C / 140 °F maximum
Boiling water	100 °C / 212 °F
Laundry iron	135 to 230 °C / 275 to 445 °F
60-watt light bulb surface	130 °C / 266 °F
60-watt light bulb filament	2500 °C / 4530 °F
Operating automobile engine	105 to 125 °C / 220 to 260 °F
Candle flame	600 to 1400 °C / 1110 to 2550 °F
Volcanic lava	700 to 1200 °C / 1300 to 2200 °F
Fireworks sparkler	1000 to1700 °C / 1800 to 3000 °F
Charcoal briquette	200 to 550 °C / 400 to 1000 °F
Propane torch	2000 °C / 3500 °F
Wood	1900 °C / 3400 °F
Acetylene torch	3500 °C / 6300 °F
Surface of sun	5000 to 6000 °C / 9000 to 11000 °F
Lightning	30000 °C / 53500 °F
Center of Sun	5000000 °C / 27000000 °F

Index

Acetone, 39
Acid, 33
Air pressure, 31
Annealing, 13

Balloon, 15,
Bathroom, 47
Bobby pin, 13
Bubbles, 45

Candle, 27
Cannon, 19
Capillary tube, 25
Chemical reaction, 33
Coin, 33
Condensation, 47
Copper, 33

Ecology, 55
Engineering, 43

Fire tornado, 41
Fleas, 37
Fluorescent bulb, 9

Gelatin, 49
Gravity, 23
Gremlins, 39
Gyroscope, 15

Hardening, 13

Inertia, 17,
Internal reflection, 21, 49

Laser pointer, 49
Leaves, 53

Metallurgy, 13
Microwave, 35
Mirror, 47

Nervous system, 29

Optic fiber, 49

Pencil, 51
Penny, 33
Piezo, 19
Pipe, 11
Polymer, 39, 51
Potato, 27

Reflection, 21
Reflex, 23
Resonance, 11,
Rope trick, 55

Safety rules, 5
Salad, 17
Salt, 43
Sandwich bag, 51
Science trick, 27, 47
Sound waves, 11
Species, 53
Static electricity, 9, 37
Stomata, 53
Styrofoam®, 39

Temperature, 29
Tempering, 13
Thermometer, 25
Tool kit, 60
Tornado, 41
Transpiration, 53
Trash bag, 31
Tree, 53

Vacuum, 31
Vortex, 41

Water vapor, 47

Zinc, 33

About the Wizards

Nearly 210 years ago, the son of a poor blacksmith read a book he was repairing while performing his duties as an apprentice in a bookbinder's shop. This book captured the imagination of the young teenager, Michael Faraday. He delighted in learning of recent discoveries in the fledgling field of science. That scientific interest led him to a position at the Royal Institution of Great Britain, where Faraday made many great scientific discoveries of his own, one of which changed our world forever. However, he is most renowned for his ability to communicate science to the public.

Michael Faraday was indeed a masterful science communicator. Queen Victoria called him "Wizard; a man who caused us to experience the truest beauty of the Creation."

Other science communicators have answered the call to be providers of common sense and understanding of scientific innovations. At the end of World War II, the term "atomic energy" entered the public lexicon, leaving most of the public fearful and apprehensive of anything tagged "nuclear." Dr. Hubert Alyea at Princeton was engaged by the United Nations to visit 88 countries, explaining the nature and positive possibilities of atomic energy to millions of people in attentive audiences. Often compared to Faraday, Dr. Alyea took the role of Wizard II, and served as the key inspiration for the Disney movie *The Absent-Minded Professor*.

In the 1950's, when the Sputnik satellite began the space race and subsequent pursuit of science learning, the public turned to yet another renowned science communicator, Don Herbert, better known as television's Mr. Wizard. For five decades, generations of youngsters were inspired to learn the "science of everyday living" from the third great Wizard.

The current holder of the Wizard title – Steven Jacobs (Wizard IV) – was mentored by two of those famed science communicators. Alyea taught him chemistry and the art of being a scientific raconteur. Mr. Wizard trained him to follow in his footsteps in television. In 1995, at the Royal Institution of Great Britain, near the memorial of Wizard I, Michael Faraday, the torch was passed. Wizard IV was given the challenge of carrying the science communicator's "light of illumination."

Perhaps you can be Wizard Five!